10301357

DISCARD

Oceans and Seas

Atlantic Ocean

John F. Prevost
ABDO Publishing Company

visit us at
www.abdopub.com

Published by ABDO Publishing Company, 4940 Viking Drive, Edina, Minnesota 55435.
Copyright © 2003 by Abdo Consulting Group, Inc. International copyrights reserved in
all countries. No part of this book may be reproduced in any form without written
permission from the publisher.

Printed in the United States.

Photo Credits: Corbis

Contributing Editors: Kate A. Conley, Kristin Van Cleaf, Kristianne E. Vieregger
Art Direction & Graphics: Neil Klinepier

Library of Congress Cataloging-in-Publication Data

Prevost, John F.
    Atlantic Ocean / by John F. Prevost.
        p. cm. -- (Oceans and seas)
    Includes index.
    Summary: Surveys the origin, geological borders, climate, water, plant and animal
life, and economic and ecological aspects of the Atlantic Ocean.
    ISBN 1-57765-092-1
    1. Oceanography--Atlantic Ocean--Juvenile literature. 2. Atlantic Ocean--Juvenile
literature. [1. Atlantic Ocean.] I. Title. II. Series: Oceans and seas (Minneapolis, Minn.)
GC481.P74 1999
551.46'1--dc21

                                                                    98-11988
                                                                        CIP
                                                                         AC

Contents

The Atlantic Ocean

The name Atlantic means "**Sea** of Atlas." Atlas ruled the legendary land of Atlantis. No one knows if this land truly existed. It may have disappeared into the ocean after an earthquake long ago. The Atlantic Ocean is named for this story.

The Atlantic Ocean is an S shape. It separates Africa and Europe from North and South America. The Arctic and Southern Oceans make up its northern and southern borders.

The Atlantic covers 30 million square miles (77 million sq km). This makes it the world's second-largest ocean. It contains about 25 percent of the world's water.

The **equator** divides the Atlantic into the North Atlantic Ocean and the South Atlantic Ocean. The Atlantic also includes many seas, gulfs, and straits. A few are the Mediterranean Sea, the Caribbean Sea, the Gulf of Mexico, the Strait of Gibraltar, and the Davis Strait.

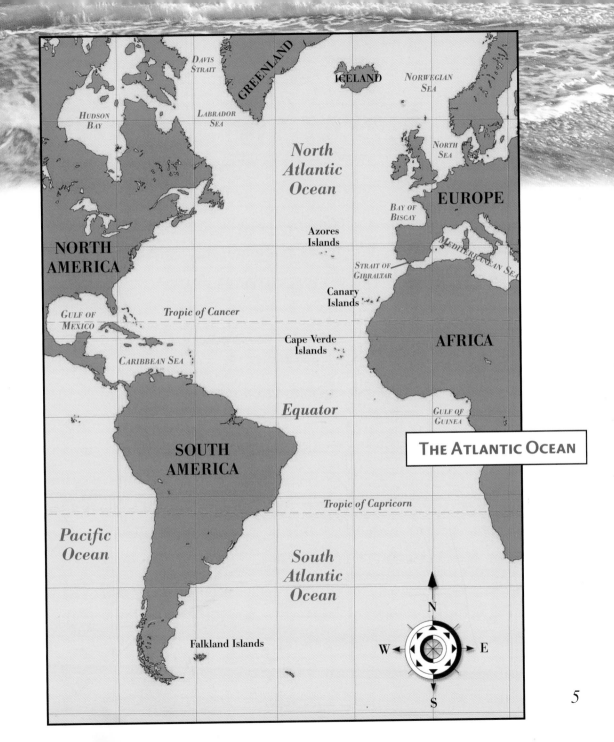

THE ATLANTIC OCEAN

Plates & Mountains

Earth's surface is broken into several large **plates**. They move very slowly. Millions of years ago, this caused Earth's one large continent, Pangaea, to slowly break apart.

As Pangaea broke, today's continents formed. They split the large ocean, Panthalassa, into today's oceans. The Atlantic is the youngest of these oceans.

Today, Earth's plates continue to move. In the Atlantic, the seafloor is spreading apart. This occurs at a large, underwater volcanic mountain range called the Mid-Atlantic Ridge.

There, two plates are slowly pulling apart. As they move, hot liquid rock called magma rises up from inside Earth. When the magma reaches the surface, the ocean water cools it. New ocean floor then forms between the plates. The Atlantic basin widens an average of one inch (three cm) each year.

Earthquakes and volcanoes also occur along the ridge. Islands continue to form from the eruptions. The island of Surtsey, near Iceland, formed from volcanic eruptions in 1963.

A river of lava pours from Mount Etna, in Sicily. It is the highest active volcano in Europe.

Land Under Water

The Mid-Atlantic Ridge is the Atlantic Ocean's main feature. It runs north and south through the center of the Atlantic's basin. It is four times longer than the Himalaya, Rocky, and Andes Mountains put together.

In some places, the ridge rises above **sea level**. The peaks form volcanic islands. The Azores, Saint Helena, and Iceland are examples of such islands in the Atlantic Ocean.

On either side of the Mid-Atlantic Ridge lie flat plains and rocky hills. They form smaller basins within the Atlantic. Ancient volcanoes are found in some of these areas. The floor gradually rises up to the continental slope, then to the **continental shelf**, and up to the continents.

The seafloor is also cut by steep, narrow canyons called trenches. Trenches are the deepest areas in the oceans. In the Atlantic, the deepest point is the Milwaukee Deep. It lies in the Puerto Rico Trench, at 27,493 feet (8,380 m) below sea level.

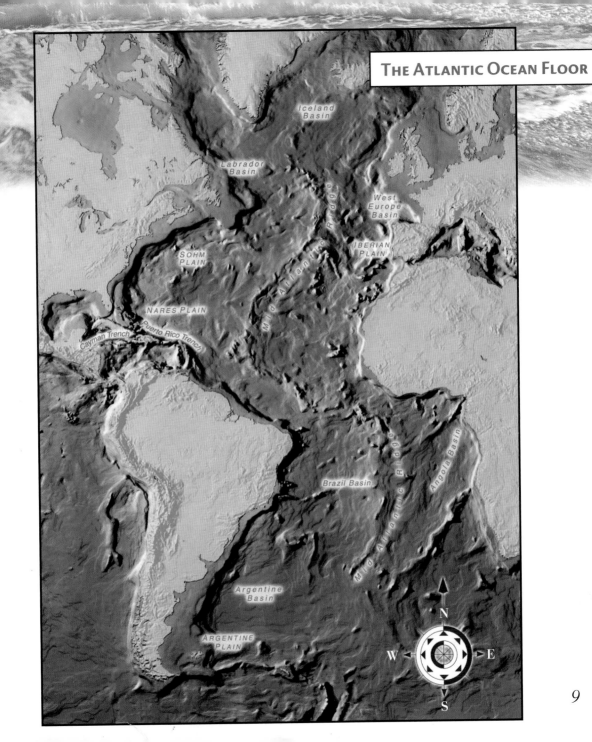

Atlantic Waters

The Atlantic Ocean's water reaches the Atlantic through rain and rivers. The Mississippi, Amazon, Rhine, and Niger are major rivers that flow into the Atlantic. The hydrologic cycle moves the water between land and the ocean.

The Atlantic's water contains salt and other **dissolved** minerals. The minerals come to the ocean through rivers and underwater **rifts** in the ocean floor. Gases, such as oxygen and nitrogen, are also found in ocean water.

Streams called surface currents run through the Atlantic. They are caused by wind, differences in temperature, and Earth's rotation. Currents north of the **equator** move clockwise. However, south of the equator, they flow counterclockwise.

Density currents move up and down. Cold, salty water is dense, so it sinks. The warmer, purer water rises to the surface.

Tides are the daily rise and fall of the ocean. The moon's gravity pulls on ocean water, creating a bulge, or high tide. At the same time, the opposite side of Earth also has high tide. Low tide occurs in between. The tides usually change twice a day, as Earth rotates.

THE HYDROLOGIC CYCLE

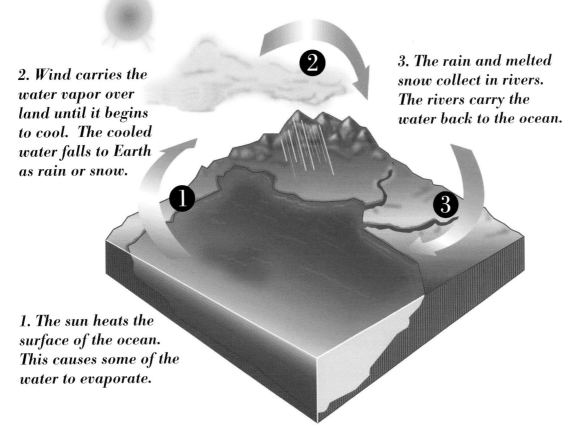

2. Wind carries the water vapor over land until it begins to cool. The cooled water falls to Earth as rain or snow.

3. The rain and melted snow collect in rivers. The rivers carry the water back to the ocean.

1. The sun heats the surface of the ocean. This causes some of the water to evaporate.

11

Atlantic Climate

Currents affect the Atlantic Ocean's climate. Currents flowing from the **equator** are warm. Cold currents flow from the poles. These currents warm or cool the land they pass near. The Gulf Stream is one of the most powerful warm-water Atlantic currents.

Winds also affect climate. In the **midlatitudes**, the westerlies blow from the west. The easterlies blow from the poles. Where these winds meet, the difference in their temperatures can cause storms called cyclones. They are most common in the North Atlantic.

Near the equator, tropical storms often occur. They form over warm ocean waters in the late summer and early fall. A tropical storm is a hurricane if its winds are faster than 74 miles per hour (119 km/h).

Icebergs are found in the far north and south Atlantic. These large chunks of fresh water break off from glaciers and float out to sea. Only their tips are visible above the water's surface.

Astronauts on the space shuttle Discovery took this picture of
Hurricane Elena on September 1, 1985. Because Hurricane Elena
formed north of the equator, it turned counterclockwise.
Hurricanes that form south of the equator turn clockwise.

Plants

Much of the plant life in the Atlantic Ocean lives by **photosynthesis**. So most plants live in the shallower waters near shore, where sunlight can reach. Others float in the ocean's surface waters.

Tall cordgrass grows in a salt marsh along the Atlantic coast in Maine's Acadia National Park.

Phytoplankton are tiny, free-floating plants. Most live in the upper layers of the water. They drift with the currents. Phytoplankton make up most of the ocean's plants. They are the base of the ocean's **food chain**.

Red and brown algae also grow in the Atlantic. Algae are also called seaweeds. They often grow on the ocean bottom, usually within 330 feet (101 m) of the surface. Dulse and laver are types of edible Atlantic seaweeds. *Sargassum,* or gulfweed, floats in the Gulf Stream and the Sargasso Sea.

Types of sea grasses also grow along the Atlantic's coasts. Eelgrass and turtle grass are two common types. They are often a source of food for marine animals.

Seaweeds cover rocks along a Caribbean shore.

Animals

The Atlantic has hundreds of animal species in and around its waters. In the ocean, animals float or swim in the water. Others live on the seafloor.

Animals that float are zooplankton. Most zooplankton are too small to see. They eat phytoplankton and are often food for larger animals. Some zooplankton are larger, such as krill, jellyfish, and Portuguese man-of-wars.

Nekton are animals that swim. They include dolphins, whales, squid, and many types of fish. Most live in the warmer, brighter upper layers of the water. But some can live in the cold darkness of the Atlantic's deeper water.

Horseshoe crabs live in shallow ocean water. They come ashore to lay their eggs.

Benthos animals live on the ocean floor. They include clams, lobsters, sea anemone, and coral. Horseshoe crabs are common on the United States's Atlantic coast. Coral **reefs** are mainly found in the warm waters of the Caribbean Sea.

Atlantic mammals include whales, sea lions, and manatees. Caribbean manatees live in tropical coastal waters. They have flippers, paddle-like tails, and eat sea grasses. They can be 8 to 15 feet (2 to 5 m) long and weigh up to 1,500 pounds (680 kg)!

Manatees spend their whole lives in the water. They are the only marine mammals that feed entirely on vegetation.

Traders & Explorers

The Atlantic Ocean has been important to humans for thousands of years. Peoples such as the Carib fished in the Atlantic. They also traveled the ocean in canoes. The ancient Phoenicians and Egyptians fished and traded as far out as the west African coast.

The Atlantic's first recorded explorations were by Viking and Norse sailors. They discovered Greenland, Iceland, and North America. In the late 1400s, the Portuguese explored the west African coast.

Many Europeans set out to find trade routes to India. Instead, they found the Americas and their peoples. Christopher Columbus sailed across the Atlantic from Europe to the Caribbean in 1492. John Cabot, Ferdinand Magellan, and other Europeans also explored these areas. The explorers drew maps, expanding people's knowledge of the world.

Today, the main Atlantic exploration is under the water. Scientists study the ocean floor and deep-sea life. They continue to study the Atlantic's climate, currents, plants, and animals.

Modern-day versions of Columbus's ships, the Niña, *the* Pinta, *and the* Santa María

The Atlantic Today

Today, people find many uses for the Atlantic Ocean. It is an important source of food. Miners drill for **petroleum** and natural gas under the **continental shelf**. Useful minerals are taken from the water. The Atlantic is also used for trade and transportation. It is one of the world's busiest shipping areas.

But human activity has damaged the ocean. Industry, mining, and garbage dumping pollutes Atlantic waters.

Chemicals and minerals from farms wash into the ocean with rivers or rain. This pollution upsets the Atlantic's **ecosystem**. This has endangered many animal species.

An oil-drilling platform near Atlantic City, New Jersey

Overfishing is also a problem. People catch so many fish that some species are endangered. Some endangered Atlantic animals include manatees, seals, sea lions, turtles, and whales.

Many countries have laws protecting loggerhead turtles and their eggs.

Glossary

continental shelf - the shallow area around each continent.

dense - having a large amount of matter in a given volume.

dissolve - to break down and spread evenly throughout a liquid.

ecosystem - a community of organisms and their environment functioning as one unit.

equator - an imaginary circle around the middle of Earth.

food chain - an arrangement of plants and animals in a community. Each plant or animal feeds on other plants or animals in a certain order. For example, phytoplankton are eaten by small fish, small fish are eaten by large fish, and large fish are eaten by humans.

midlatitudes - the warm regions just north and south of the equator.

petroleum - a thick, yellowish-black oil. It is the source of gasoline.

photosynthesis - the process by which green plants use light energy, carbon dioxide, and water to make food and oxygen.

plates - large, moving sections of Earth's surface on which the planet's continents and oceans rest.

reef - a chain of rocks or coral, or a ridge of sand, at or near the water's surface.

rift - a long, deep, narrow crack.

sea - a body of water that is smaller than an ocean and is almost completely surrounded by land.

sea level - the average height of all the oceans. It is often used to measure height or depth, with sea level as zero feet.

How Do You Say That?

algae - AL-jee
anemone - uh-NEH-muh-nee
cyclone - SI-klohne
hydrologic - hi-druh-LAH-jihk
Pangaea - pan-JEE-uh

Panthalassa - pan-THA-luh-suh
Phoenicians - fih-NEE-shunz
photosynthesis - foh-toh-SIN-thuh-suhs
phytoplankton - fi-toh-PLANGK-tuhn
zooplankton - zoh-uh-PLANGK-tuhn

Web Sites

Would you like to learn more about the Atlantic Ocean? Please visit **www.abdopub.com** to find up-to-date Web site links about the Atlantic Ocean, its islands, and its climate. These links are routinely monitored and updated to provide the most current information available.

23

Index